THE FLATLAND DIALOGUES

David Sayre

ISBN: 978-1-937721-40-4

Library of Congress Control Number: 2017955800

Published by
Peter E. Randall Publisher
5 Greenleaf Woods Drive, #102
Portsmouth, NH 03801

davidsayre.com

Preface

Everyone wonders if there's more to life, even scientists.

But as science advances, our traditional notions of heaven and salvation retreat. Must all hope be abandoned, as we reach farther into our understanding of the universe?

As we learn to modify our biology, we turn the corner of evolution. Now it is in our hands, for good or ill.

Might science support a broader view of life? Might there be more to life than is apparent to us?

Rebecca and I wrote the picture-book *Flatland** to give children a view of that "more." The story grew out of a talk at my own daughter's funeral. But it is a rational story, based in truth. It is real.

The *Flatland Dialogues* are about that reality.

Are we alone? What is intelligent life, really? In what full reality do we live?

Owuza Flatlander says we are not alone. Life doesn't stop at the edges of Flatland.

David Sayre

---

* Rebecca Emberley and David Sayre, *Flatland*, Two Little Birds Books, 2014; recommended by Parents' Choice Foundation

Prologue

I was in prison when I first saw Flatland.

Doc Savage was there, and the Boston Strangler, and a handsome, articulate inmate who had chewed out an opponent's throat in a fight. Strangely I felt safe among them, and they were full of advice.

> *We failed at everything, Dave, even crime.*
> *We're all numbers in here, Dave, got no names.*
> *We all got walls inside us, Dave, higher than those.*
> **This is Flatland, Dave. You can't look up. Just side to side, front and back.**

We all walk on Flatland. Going back and forth between my laboratory and the prison, I began to see what makes it.

I was developing satellite communications at one end of that road and teaching electronics to prisoners at the other. We won some patents and graduated some inmates. But the important discovery was what makes Flatland. Why is our world full of edges and endings? Why can't we look up?

The answer lives especially in laboratories and prisons, and it's the only thing on which I'm actually expert. We call it "entropy." I can show you its formulas, but it boils down to disintegration and ignorance.

Intelligent life is all about reducing entropy.

So the Search for Extraterrestrial Intelligence—the SETI Project, founded by Carl Sagan—gives us a good start. They're on my cover, talking to my principal character, Owuza Flatlander. Their reach is our reach, and it is beautiful as well as rational.

These *Dialogues* explore Carl's question, starting beneath a radio telescope in a remote SETI outpost: Are we alone?

David Sayre

Contents

BOOK I

SETI Outpost, January 1.

The search for extraterrestrial intelligence continues on Earth. In this remote outpost, scientists listen for some evidence that we are not alone. Here the searchers (S) engage in dialogue with Owuza Flatlander:

S: My people see something of their story in *Flatland*. They have a sense that there's more to life than is apparent to them.

> **O:** There surely is. And who are your people?

S: The skeptical, the student. Those who seek comfort, who seek meaning in their lives, who seek direction. Those in grief, in pain…And your story seems to have several levels of meaning. Is it intended for children, or for their parents, or others?

> **O:** Yes! *Flatland* is a children's picture-book that offers comfort and some level of meaning to kids who have lost someone very close to them. But for an older reader—or a parent struggling to understand, to explain—it's very important to be honest, don't you think?

S: Sure is. Our stories have to be rooted in truth, not some fantasy that the parent doesn't really believe. Not some story that the child later realizes we just made up to make them feel better. But is it really possible to be both honest and comforting?

> **O:** Yes, of course. That's the point of our story. Flatland is two-dimensional. Even the child realizes that the Flatlanders are missing something. We don't look up; we can't see off our page. There's more to reality than we first perceive.

There's more to each other, more to relation. There's more to ourselves! That's because our senses evolved just to survive in Flatland! Is that true also for humans on Planet Earth?

S: It's true that our senses and brains evolved to survive on Earth, not necessarily to perceive all of reality. That's why we're searching for life beyond Earth…But we don't have any patience for fuzzy thinking and we don't expect to find a Heaven. Your authors have drawn some evocative pictures, and we think there may be some new ideas in there, but they have to be real. So, where do you think we should be looking for extraterrestrial intelligence?

O: You've been looking for organic life-forms like those found on Earth. So, you've detected many candidate planets around your universe, where liquid water might be found. But I think you realize that intelligence may not be limited to familiar organic forms. How would you know when you've detected it? By a wave or a smile? A noise? A smell? Perhaps an attack?

S: Of course we know that's silly. We talk about other "earths" and "life forms" because people can understand that. And frankly we need public funding to keep searching. Of course we realize that other intelligent beings probably haven't evolved the same as we, and may not use the same means of communicating…They may even be all over, and we just can't detect each other…No, what we're really looking for is some kind of non-random signal. Something that wouldn't happen accidentally, or that isn't just endlessly repeating. Some evidence of an ability to recognize order and the freedom to choose it. It takes intelligence to choose orderly patterns and communicate, right?

O: Exactly so. The recognition of order, the freedom to choose it, the ability to communicate—those are some of the ways intelligence is expressed. What would you say they have in common?

S: That's easy. They all reduce what we call "entropy." We experience entropy in two forms, you know—the disintegration of everything unless you build it up, and the information content in any message until you receive and decode it. Decay and ignorance, if you will. They're mathematically equivalent, you know. You can reduce entropy in a given system at the expense of a greater increase in the surroundings…well anyway, intelligence is what lets us put things together out of chaos, and detect messages out of noise. That's what we're looking for.

O: And where would you say these things might come from?

S: I would say from the physical and mathematical laws that are invariant around the universe. We believe in a rational universe based on those laws. The universe is probabilistic, but not capricious. By that, I mean there's a reason for everything that happens, even though we may not be able to see it. We don't believe in miracles…I hope you're not going to tell me that there are mystical wonders or things we have to take "on faith"—?

O: Quite the contrary. It's that rational universe that supports the things we care about. Werner Heisenberg, your great quantum physicist, called what you're describing the Central Order. That invariant reality exists of itself. And what other experiences reduce entropy among intelligent beings?

S: Well, given the freedom to choose order and communicate, organization is possible, even likely. And that leads to learning and building things together…and I'd say eventually healing each other, and so forth. So, an intelligent society or association of some sort could survive that way, could keep reducing entropy locally. And there are probably a lot of them that we haven't yet found, right?

> *O:* Yes, of course. And to survive, they have to develop a farther reach in each other—do you know what I mean?

S: I know what you're getting at. Things like love and beauty are important to us scientists, too. But we have to explain them rationally, in the context of reducing entropy. So, take love, for example. It can be seen as a combination of commitment and trust and a kind of "extension" to each other. By that, I mean what the psychologists call "identifying"—empathy, sharing experience, feeling like part of each other. It's most intense between two persons, but a society needs that kind of stretch to each other, to reduce disintegration and ignorance, to survive. So, I'll grant that love is real in that sense, and essential...Or did you have some more spiritual explanation in mind?

> *O:* No, I like your explanation. Love is an expression of intelligence. To be real, it must be rational, that is, faithful to truth. And what about beauty—that may be the hardest phenomenon to explain, don't you think?

S: I've thought about that, too. I know many folks like to think beauty is beyond explanation, is mystical, some kind of spiritual experience that we shouldn't try to rationalize. But to me, seeking to understand anything is the greatest respect we can pay it. If beauty is real, if we hope to find it among all intelligent beings anywhere, we have to be able to say what it is, to communicate about it. Would you agree?

> *O:* I would. So "beauty" should not be based on some set of rules or standards, right? Nor should it be considered beyond our ability to understand. And like love, it would not be contained in what seems its source. Yet it's not a law of itself. So it must be a capacity of intelligence. You would

expect to find it wherever you find intelligence expressed, would you not?

S: Yes, it's almost like an instinct to us. Over and over, discoveries have come from an expectation that an "elegant" solution should exist. By that we mean an understanding, often in mathematical terms, that captures a lot of information in a simple expression. Think of Einstein's famous formula for the equivalence of energy and mass, or Claude Shannon's formula for the information content in a message—just brief little formulas that say so much! And when we find it, there's a thrill, a sense of wonder. The astrophysicist Chandrasekhar is often quoted as "shuddering before the beautiful . . . [that] a discovery motivated by a search after the beautiful in mathematics should find its exact replica in Nature . . ." This seems possible only in an orderly universe, only where we all have a direct route to the same self-existing truth, as Roger Penrose likes to say. Are you with me?

   O: I am. But mathematics and nature are not the only routes . . .

S: I know, I was getting to that. We scientists often have a special reverence for the arts, for the same reasons. They evoke patterns or meaning in what would otherwise be chaotic. And at the same time they communicate intensely about that meaning. In other words, they reduce both forms of entropy! Do you know why I consider that so important?

   O: It's important because that evokes all the capacities of intelligent life. The freedom to organize, to learn and communicate truth, to heal and to love, all involve a choice of order and relation. They all reduce disintegration or ignorance—the two ways we experience entropy. So your trust in beauty reflects your faith in an orderly universe, in the shared experience of being in touch with a single truth. Is that what you had in mind?

S: Yes. So to me, beauty is a startling connection with each other. It should be a way to express intelligence more fully, to experience what's true, to share it. Therefore beauty is rational, not mystical. Well, I get a lot of arguments about that, but how do you feel?

> *O:* I agree you've found a truthful and respectful way to think about love and beauty, along with the other expressions of intelligence. The important thing is that they are real. They can be understood. They are universal. They're not things invented by Flatlanders. So, as expressions of intelligence, they can be found anywhere, would you agree?

S: We would expect those to be universal capabilities. We would expect to find them in other intelligent life-forms. That doesn't mean they exist on their own, like the laws and order of the universe. They are functions of intelligence; they need intelligence to be expressed. I hope you're not going to tell me that beauty and love and freedom and so forth are some kind of Platonic reality that exist independent of intelligent beings—?

> *O:* No, if your people are to hold something sacred, it should be intelligent life, and truth. It should be their capacity to experience those things, to choose them. The sacred thing about beauty is that we can see it, and even make it. The sacred thing about love is that we can experience it, and even give it. Those wonderful capacities are freely given to us. That freedom, that seeing and making and giving, are realities of intelligent life, as you have recognized it. They exist in our expression of intelligence, of truth, of the Central Order.

S: So, your *Flatland* story made us think about how we might explain them—and whether they give us a rational basis for moral direction, especially as our old religious explanations lose their grip. So do they?

O: Yes, they do. Being rational just means being faithful to truth. That is your commitment as scientists. We Flat-landers couldn't accept some miraculous revelation either, or some set of rules, as a guide for living—or comfort in dying. We had to find the reality, the meaning, that wasn't obvious to Flatland senses. Rational thinking is an expression of intelligence. We knew the "more" we were seeking had to be there. Isn't that what you're seeking?

S: Well, science is about discovering truth, and our mission is to discover other expressions of intelligence. We'd like to tell our children that learning to help and commit to each other is more than just a set of rules. *Flatland* got us thinking about, well, "looking up"—maybe there's more to our search for intelligent life than whether we're alone in the universe. But we can't program our telescopes and probes to detect something else, can we?

O: Actually, that's what brings me here. Your people may survive. Many civilizations don't. You're reaching for intelligence itself. You just told me how you'd recognize it. And another thing—you've figured out how to use phased arrays. Do you know what I mean?

S: No...I mean, our arrays of telescopes and probes combine different points of view coherently, so we can see a smaller target at a greater distance. What does that have to do with our survival, or our moral direction?

O: You may have noticed there are some general principles at work there. You're using a great diversity of viewpoints. From each one you're extracting a small, true signal out of a great, loud noise. Then you're combining those truthful perceptions "in phase"—in cooperative perception. This is not just metaphor; it's how civilizations survive. It's how isolated individuals can choose to commit to each other and to truth. Do you see the principles?

S: I see the value of diversity in combining viewpoints. And being truthful, of course. And working together like, in phase. I get the part about survival as a species depending on those commitments. Those things should work wherever there's intelligence. Without them we eventually blow each other up, or poison our nest…Now I'm worried that, maybe that's why we've never detected intelligence anywhere else, at least so far—maybe they all kill each other. Or maybe they see us doing it, and they're hiding. Do we have any hope?

O: That's why we're talking. Your Darwinian evolution has run its course. Evolution on Earth is now in your hands. Now you have a choice. That's where your hope lies, and of course your risk. You've just told me the general principles of your direction. And you've recognized that you're searching for intelligence itself beyond Earth, not just for the forms it might take. Do you see in your mission a way to steer your evolution forward?

S: Well, we know evolution started with the self-assembly of organic compounds, which are common in our universe. Some reacted accidentally with each other to form membranes and nucleic acids, joining in the first archaic cells. From there bacteria could develop and thrive, maybe a billion years after the earth formed. Multi-cellular structures formed about two billion years ago, and evolved by natural selection into plants and animals like us. Our brains grew by their selective advantage since then…well, anyway, so what?

O: So gradually instinct has become choice. Gradually the degree of your freedom has grown. Your ability to organize and communicate has increased. Now you can choose to heal each other, to learn and build together. You can see and make what you call "beauty," which reduces both forms of entropy. Finally, you can make commitments and trust and extend yourselves to others, which you call "love." That's an entirely logical direction, a rational moral compass. It's based simply on truth and the reduction of ignorance and disintegration. Can you now build those capacities without limit?

S: I'm sure we could, given enough time and cooperation. But not everyone believes in those things or will work together. The world is full of ignorance and short-term self-interest. How can we get over that?

O: That's your job, isn't it? To reduce ignorance and disintegration, the two forms of entropy. As scientists searching for intelligence, you have the mission, the platform, the voice…So tell your people *Yes*, we have found evidence of intelligent life. *Yes*, it can be anywhere. Intelligence is built into the universe. It doesn't need us to exist. It is not based on mystery but on the Central Order of the universe and its invariant laws. It's probably expressed in many forms, perhaps an infinite number of forms, most of which we can't yet see. But we do know *how* it's expressed, how we'd recognize it. Intelligent life is seen in the ability and freedom to learn and build and communicate truthfully, to heal each other, to recognize beauty, to make commitments and trust and extend one's self to others. These are universal expressions of intelligence. Exercising them is a capacity given to us all freely. In choosing them together, we live. Can you spread that message?

S: I agree it's the right moral direction. And it gives us a sense of meaning, a way to live. It's logical and understandable, so we scientists could subscribe to it. However, if we want others to accept it, it has to convey comfort somehow, some hope of self-interest. Their old religions assured them of a God who would take care of them in Heaven. How do we answer the "What's in it for me?" question?

> *O:* The answer to that question lies in what you mean by "me." Of course, the individual, isolated packages disintegrate, and meanwhile are subject to all manner of ill fortune and attack. They change all the time, there's no fixed individual. Perhaps you will agree there's nothing sacred about the evolution of such packages from the mud of Earth. In fact, you will eventually duplicate that evolution in your laboratories. Even sooner you will manipulate your genetic inheritance, modify your species. Is that the "me" you care about?

S: I suppose, at least as we mature, what we really care about in each other—and I guess even in ourselves—is not that "package" subject to entropy, but those qualities that reduce entropy. I mean, for ourselves and those around us—those relationships and behaviors and commitments that we've been talking about are what count. That's how we really express intelligence. Well, to the extent it's truthful and beneficial anyway—that's what we really care about. That's the person I love, if you will. Okay, those things don't die themselves, but how does that help the person trying to practice them faithfully, especially when they're in pain of some sort?

> *O:* To be dependably comforting, that larger concept of a "person" has to be real to them. They must realize that their whole "me" includes those expressions of an unbounded intelligence. They're not stuck inside a disintegrating package. In fact, they integrate—add to, become part of, are included in—that central reality as they express its qualities truthfully and in benefit to each other. Those

entropy-reducing commitments and behaviors express a reality not limited to the boundaries of an isolated individual. Being that lasting "person" you care about is made possible by inclusion in the whole expression of intelligent life. That's the lesson of Flatland. They must see that sharing of intelligent life as their extension, their lasting identity— the "more" of themselves. They don't have to earn it, but they do have to choose it.

This is not just an opinion or another mystical creed. This is a choice made possible by the universal reality of intelligent life. In that choice, they can feel included. Can you offer that comfort?

S: Hmm. They'll recognize these entropy-reducing expressions of intelligence—freedom and learning and healing and so forth—as the better parts of their nature. Nothing so new in that. They'll be glad to hear those have a scientific basis, other than their religious teachings. That's good. I guess they'll understand the logical importance of diversity, that all these important qualities are enhanced in sharing, in combining our efforts and insights. That helps. Maybe they'll see that what we really care about in each other, as we mature anyway, is those qualities that last, not the perishable physical appearances or pleasures. They might even agree that those qualities are universal expressions of intelligence, not our invention. Still, as individuals they'll consider all that as merely like their own personalities, something which can't be defended against accident or attack, and will die with them. Where's the inclusion in something "more," that they can reach?

O: You must free them from their Flatland thinking. That part that hurts—that pain, that grief, that loneliness, that emptiness of meaning and direction—is not all of them. It's not the whole story. All that painful experience comes from a kind of "noise" reaching the brain over time, masking for a time their perception of the connections, the inclusion, that we've been discussing. That time will pass. All of experience is perception. The "person" whom you said you care about is not bound up in those pains, not contained or trapped. The qualities you care about, by definition actually reduce containment. Is that not so?

S: I suppose so, but then where is the continuity of the individual?

O: That question is still looking for some kind of distinct, separate "soul" inside each individual. You won't find it. For the isolated individual, "continuity" is a function of his or her time—in your case, Earth-time, the time that started with your universe, now in its 14.7 billionth year. If you want continuity of an isolated individual, you have to take entropy along with it—the edges and endings of Flatland. If, on the other hand, you are committed to the true expressions of intelligence among us, would you expect to find them continuous in any one individual?

S: Well no, I would say they vary all the time. We sleep, we get angry, we're easily confused, we're often selfish…Some days I learn a lot and share it well, and some days I forget, or hide things for my own career…I mean, intelligent beings vary enormously in their expression of freedom, learning, communication, and so forth—not to mention their commitments to each other. Where does that leave my hope of being more, or finding more, than what we see with our evolved senses?

O: It leaves your hope where it can be fulfilled, in each other. Not only in the separate individuals you see in your Flatland, but in the "Whole Each Other." That is, in the integration of all expressions of intelligence that meet the tests of truth and mutual benefit. By "benefit," we mean enhancing these lasting life qualities in each other. So, the "continuity" your people hope for lies is relation, not in isolation. Have you experienced such continuity?

S: Logically, the invariant physical and mathematical laws we experience in the universe are continuous realities. And truth itself is continuous by definition—that which prevails in every event. We see a continuous over-all order in the universe, which we experience as expressions of intelligence. So, we do experience space-time continuity in that sense, being devoted to those realities. But I can experience those realities alone, without what you call "relation," can I not?

O: Yes, from time to time and place to place, along your universe's space-time fabric. You are free to choose them, a key to intelligence. But you were hoping for continuity. That requires relation among all expressions of intelligence. Like your phased arrays, such expressions add coherently to the whole reality of intelligent life. You cannot hold on to a continuous, isolated, individual, exclusive identity in yourself or in those you love; but you can add to the Whole of which you are part. So, you can find yourself, and them, in relation. In others; that is, in the whole expression of intelligence. Can you explain that concept to your people?

S: How can anyone be part of others?

O: You have already answered that question. You described how intelligence is expressed. You described what you care about in others, what makes up their lasting nature. You described what is universal among intelligent beings, wherever you may find them. When you asked your "how can anyone" question, you had in mind the wrong picture. We're not talking about somehow melding the brains that evolved on Earth. This is not magic, not mystery. We're talking about the reality of intelligence, and how it's expressed universally.

In those expressions, the freedom to join, to share, to be entangled in a single system, is perfectly natural. So, finding the "more" of one's self in relation is simply through the means we've been discussing—by healing, commitment and trust, by learning and organizing to mutual benefit, by communicating truth, by seeing and making beauty, by setting each other free. Those are the means of relation, among the life-expressions you care about.

S: Yeah, but folks will say those things die with the individual. They all require a brain, and brains can be fooled or injured or manipulated.

O: Those things were never contained in an individual. Nor do they require brains to exist. That's the point. Intelligence is built into the universe, in the single system of invariant laws and order that you have described. Its expressions take many forms, probably an infinite number. These expressions come and go, start and stop, along their space-time dimensions. Some would look like a brain to you, others evolve different forms. They may appear small and transient, or larger and more lasting. Would you expect intelligence to be found only within the boundaries of life-forms like those on Earth, or on Flatland?

S: Probably not. I remember Fred Hoyle wrote a novel about a "black cloud" of energetic particles that had organized into an intelligent being. It only takes freedom to organize and exchange information for intelligence to be expressed anywhere. That's what we hope to detect. But I wouldn't expect any beings we can see to live forever…

O: They would live in relation, not in isolation. Their essential life-qualities are shared, not owned. They express a reality that exists of itself, not a Flatland or human or "alien" invention. Logically, they would include all truthful and beneficial expressions of freedom, communication, learning and building, healing and loving, perceiving and making beauty. Many products of evolution or fabrication might express those life-qualities. But as part of a single system, they can be considered "entangled," as you have said about quantum particles not subject to space-time confinement. Your people do not really want to live alone. They want to live in relation. They want to find meaning and direction in choices that are real and universal, do they not?

S: Oh, when they think about it, I suppose. Even we scientists don't contemplate such things often. And anyway, what could I call this integration of intelligence-expressions? I mean, this is a long discussion of abstractions that folks don't regularly think about—?

O: Actually, they do think about it regularly, but their tradition is a supernatural God. They think about miracles, and mysteries, and angels. But this is real. So, might you call it the Whole Each Other? The reality in which we live?

S: I don't know. This is all logical, but it seems so far from everyday experience...

O: That's a result of thinking from your ground up, from the immediate experience backwards. But you know there are dimensions of reality not evident to senses evolved by natural selection on a particular planet. You work with them all the time—the invariant laws you have discovered, the fundamental forces of nature, quantum particles and fields, the underlying order of the universe. They also seem remote from everyday experience, but they make it all possible. Order is a product of intelligence. Isn't it logical to understand intelligence as you understand those other realities, as a reality of itself, a single, pervasive potential?

S: Yes, it's logical, but I can't prove it.

> *O:* You don't have proof, but you have plenty of evidence. We know how to recognize expressions of intelligence, which are all around us but not all evident to Flatland senses. The forms they might appear in, along their own paths of evolution, could vary enormously. Each might consider their forms the only ones. But no matter their form, each could express intelligence truthfully, in the ways we've been discussing. That universal reality, that Central Order, is what makes their choice possible. In that choice, though inconstant, they commit and extend themselves to the whole of intelligent life. In each such expression, however inconstant, they are integrated in the Whole, they add to it. You understand coherent integration. Can you help your people understand that integration, in the greater reality you're devoted to?

S: Well, we can offer it, and defend it I guess, as a faith rooted in truth. It's a direction that has some hope of survival for all of us. But folks don't feel any different when making those commitments, they don't sense a new "connection"—and anyway are we asking them to give up their individuality, which they've always fiercely defended?

> *O:* Of course, the Flatlander evolved—and so is entirely conditioned to—a self-image as a separate individual, with an exclusive continuity of being. Yet that view is, as your Einstein once said, "a kind of prison for us." It is that conditioning to isolation, that apparent boundary, that exclusion, that limits the life-expression. What we're talking about is the freedom to choose connection, to see one's self in a wider reality, to extend one's self to others. Surely they can understand that they have adopted a Flatland view, it's obviously not all of reality—?

S: We can make that case, but how do we make that wider reality real to them?

> *O:* You have already discovered its elements in your own work. You have described an inclusive reality that exists of itself—a discoverable truth, including invariant laws—a non-capricious, orderly multiverse. The achievement of that order is how we've described intelligence itself. That intelligence is a universal potential that exists of itself. Your people express it, as they choose relationships and behaviors that reduce entropy around them in the same way. That is by definition a sharing of intelligent life, an inclusion in it...

S: ...and that expression is what reduces entropy among us—it actually reduces ignorance and disintegration, at least for that system. So, it doesn't die with the package. Is that what you're driving at?

> *O:* Yes, and these expressions add to the Whole and are part of it. Did you study Rebecca's drawings in our book?

S: Actually, I did. The way she showed Flatlanders and the universe around them, well, she's saying, we're all "made of the same stuff." To the extent we express those true life-qualities faithfully, anyway. That's what I liked about the book. So look, I'm confident that truth exists on its own. There's an order in the universe that we can count on. Intelligence lets us increase the order around us. The experience of love and healing, as we've defined them here, is real. Beauty and learning are real. I think most scientists will agree with those premises, once we rationalize them. Rebecca's "inclusion" concept, though, has to be real, rational, not mystical. How do we rationalize that?

> *O:* You have reasoned that out from first principles. The inclusion escaped us Flatlanders too, while we were focused on a continuity of our "selves" in Flatland. But then we realized that we changed all the time. In form, of course, from day to

day, but also in our expression of the things that matter. We are truthful and helpful sometimes but not always. We had to realize that this fundamental expression of the life-qualities is not time-limited, so we had to stop thinking about a continuous, fixed, separate, isolated "soul" to defend. That's why Rebecca drew the images in our book all "made of the same stuff." The nature of intelligence is integrative. Our expression of the life-qualities isn't continuous when it's isolated. When we looked up, we could see that we lived in each other, in our truthful and beneficial expressions of intelligence. That's what our Flatland view obscured.

That capacity to love connects us. That choice of commitment to others is an expression of a universal reality. The capacity to extend to each other expresses an underlying wholeness. Making that choice adds to the Whole. As you have established, the nature of one's being that we really care about is her expression of those life-qualities, not the form in which each evolved (or fabricated) being is seen by Flatland senses. Those qualities—that expression of life—is made possible by the primordial laws of the Central Order, the universal intelligence we've been discussing. That expression does not depend on Flatland forms, is not contained in them. The expression of intelligent life is not subject to the time that was born with our universe, not bounded by Flatland perceptions.

S: That's an awfully long answer. How can we make it into something our people can grasp?

    O: That's why we wrote *Flatland.* Your people don't have to go through some intellectual exercise to experience this reality. Their self-interest is found in others, as part of a "Whole Each Other," that is, of all intelligent life. It's this Whole—this Central Order of the universe, this intelligence potential—that really is the source of all these capacities of life, as you've recognized. The potential of connection is the "more" of each one, and of relation. They're given the means of that connection freely. Isn't that a rational faith your people could embrace?

S: They'll have a hard time giving up their supernatural gods in exchange for some Central Order that only we scientists understand. And they evolved to compete for resources, not to give up individuality. They don't even like each other much of the time. They won't see any self-interest in a "Whole Each Other," or how to be part of it. Besides, there's always this "my god is more powerful than your god," and so forth. How can they put their trust in such a different direction?

O: It's not so different, just more universal, and more real. They can choose isolation or inclusion. Freedom and learning and sharing are real, rooted in the primordial intelligence that's as real as the invariant laws you've found. It should be easier to trust in one's inclusion in a universal intelligence, than in miracles and heaven and hell. Look, your people need a source of moral direction, of meaning and purpose, and of comfort. Their old religions tried to supply those needs. Folks don't have to give up that aspiration, just recognize its more lasting source. One cannot command success, but one can always be faithful. Don't you think they're ready for that?

S: Some of them, I suppose. Most of them I'm afraid will want to hang on to their salvation and creation myths. Where else can they find a sense of being loved by God, or forgiven for sins, and so forth?

O: Oh, that's the good news. All those things can be found in a better place than mystery, and are more dependable.

Your people pray for what they consider the "love of God." Reality is better than that. You've recognized what we really mean by love. So, we are given a perfect *commitment,* in the invariant Central Order of the multiverse and the universal potential of intelligence; a perfect "trust" in our ability to be faithful each new day, no matter our past; and a perfect "extension," that is our inclusion in the whole of intelligent life. Aren't those the elements of love as a rational reality?

S: Sure, we've agreed that's a logical meaning of "love." I'd say it's also what folks mean by "faith." But will they accept that kind of faith as they lose it in their old religions?

O: You do have a head start. Your people do have a sense of being part of a larger reality than they see casually around them. As they argue about it—or abandon it in despair—you have a chance to hold up for them a better reality. They know your commitment, what you're searching for. They know you can now steer evolution, for good or ill, for the first time. You have the platform. They will listen to what you've found.

S: So, must we scientists be the prophets of our day?

O: The hope of your people lies in their commitment to each other and to truth. You understand that direction. You have the means to heal and improve this species. Who else could take on this mission?

S: All right, let's say I understand the position we're in. I agree our profession is based on a commitment to truth. And I understand the principles we've been discussing. But non-scientists will want something more personal, something, well, warmer and more comforting to believe in. Our children want someone they can pray to, some kind of personal relationship…

O: Yes, that relationship is what prayer's about. A rational prayer would not seek a miracle, a suspension of physical laws. We depend on those laws, that Central Order. In fact we want to embrace it, not seek exceptions to it. We want to extend our true selves to its full reality, to reach beyond our Flatland experience. That extension, that trust and commitment, is how we've described "faith"—and it's how we really connect with those we love, is it not?

S: Well yes, but can our children "love" this Central Order, this Whole Each Other, this universal Intelligence?

O: Of course. Your people often speak of the "love of God." In your Western tradition, they might say "I love Thee" as their prayer. In expressing that central reality truthfully and in benefit to each other, they are in fact included in it. That is a perfect relationship.

S: So can we tell them that such a prayer might be answered?

O: Their answer is the unconditional commitment, trust, and inclusion freely given them by the reality in which they live. We Flatlanders also needed an understanding of being "included" in a universal intelligence. We realized that was just like the experience of love, only it extended through and beyond our two-dimensional world. That's what makes the things we hold sacred real. We found it by reaching to each other—a "Whole Each Other," that is, all truthful and beneficial expressions of intelligent life. That's the reality in which we live, and it's timeless. Is that comforting?

S: No . . . I mean, not enough. They fear death, loss, pain. How can they find comfort in this new kind of faith?

O: Flatlanders had the same problem. Looking along the flat dimensions of our space-time, where we evolved by competitive selection, each one felt isolated. But all of experience is perception. We could look up. We could find the wider reality. We had the means—by giving benefit to others, we found ourselves in them. That changed our experience, relieved our pains, assuaged our grief. We realized— as you have—that the things we really cared about in each other were not those evolved forms at all, not the things painted across Flatland. What we really cared about were those shared realities that evidenced intelligence among us. In the end, those are what we love in each other—the expression each makes of intelligence, in sharing and healing and loving. All those expressions of intelligence were enhanced, the more we came together as a whole. This is not just a way to live; it's a way to life. Can you see that self-interest can be found in other-interest?

S: Well, as a general principle, or for the whole civilization, sure. But what do I say to the mother who has lost a child, or to one in the pain of death?

O: That's the lesson of *Flatland*. The world seems to shrink down to a tiny trap in such pain. But that's not the whole story. Flatlanders don't see all of themselves or all of each other. If one's expression of intelligence is not entirely *contained* in the mechanisms that evolved on Earth—or that you may fabricate or engineer—then it is not lost when the mechanism yields to entropy. One's expression of intelligent life exists not in isolation but in relation. You must stop focusing on the mechanism, whether evolved or fabricated or edited, as an object of continuity. It may die. But it is the relation that matters.

S: How can I make that "relation" real to my people?

O: You must speak of the "Whole Each Other" in which they live. Say this: To the extent you express these qualities of intelligent life in truth and benefit, you live. You are given, from the Central Order of the multiverse, the freedom to choose truth and beneficial connection. The means are clear, and they are real: the things we've been talking about all reduce entropy, they integrate the lasting expressions of intelligence. The things we've held sacred are real, after all.

This central reality that you can express does not need your mechanism to exist, whether that mechanism evolved or was fabricated or modified. That reality includes all truthful and integrative expressions, and is not limited by the time that was born with your universe. Your additions to it, by the means we've discussed, increase the whole and cannot be removed from it.

Those expressions are what you really love in each other—your love is not for the isolated mechanisms that change daily and finally die, not for the separated forms visible with evolved senses along the edges of Flatland, but for these permanently-integrative expressions that you share. In them you are "entangled" in the single system of unlimited intelligence. In them you live, as do those you love, in that "Whole Each Other." So, find yourself, the "more" of yourself, in others, in the integrated whole of intelligent life.

S: But some will ask why we believe in the existence of intelligence other than in our own brains—?

O: You have already answered that question. Why are there physical laws we can count on? For that matter, why is there something, instead of nothing? Why is that something orderly, not capricious? The achieving of order is by definition the reduction of entropy. We know that entropy—information—grows without limit, is not "conserved." The receipt and use of information, and the intentional organization of chaotic systems, reduce entropy in any given system (at the expense of its surrounding increase). We call that process "intelligence." That order, that intelligence, those laws, exist on their own; they don't need us to exist; they are universal and primordial. Thus, the capacity to detect that order and choose it can be expressed, manifested, evidenced, in any space-time event. That capacity, to reduce both informational and thermodynamic entropy—to learn, to manipulate one's environment, and think abstractly—we recognize as our expression of intelligence. To think that our brains are necessary for intelligence to be real is a Flatland conceit.

Your people can think of this Central Order, this universal intelligence, as a kind of "field," like the Higgs field that confers mass, or a gravitational field, or an electromagnetic field full of information to be detected. They can choose "resonance" with that field. Or perhaps like invariant truth, which can be found and would be the same for all even though perceptions vary. Those are familiar experiences, with rational foundations. And they have a logical extension, in the freedom and capacity to learn and communicate and build, to heal each other, to trust and make commitments and extend ourselves to each other, to see and make beauty. Those are the means of living in others and in the Whole. They are real, whether we express them or not.

The Flatlanders couldn't hold on to separate, isolated lives that were continuous. When we learned to communicate truth, to free each other and heal each other, to make commitments and trust and extend to each other, we discovered another dimension to reality. We realized it wasn't bounded in our Flatland brains or in Flatland time. The more we shared, the more we grew.

We discovered that we were part of a reality that didn't stop at our edges. The benefit of each one was found to be in that Whole of truth and intelligent life. One could elect to remain self-interested and isolated, which is to die. Or one could seek the interest of all others and join the Whole, which is to live.

S: But probability is always on the side of entropy. Isn't this an uphill struggle?

> *O:* Flatlanders found a way to sustain the improbable despite the downward slope of entropy. That's not the only invariant law. In this Central Order of the universe, we are given means to join, to stretch. In that way, we found there was more to each other. Where would you expect to find the greatest order, the least entropy, the most improbable thing in the universe?

S: Intelligent life, I guess. It does the unlikeliest things, always pushing away entropy. That's why we're looking for evidences of intentional order—highly improbable signs of organization. That should be the direction of evolution anywhere, isn't it?

> *O:* Yes, of course. The story of evolution is one of growing together, of mutual benefit, of finding truth. That process is not finished, on Earth or elsewhere. Despite examples of destruction and ignorance, you can see great and accelerating progress in healing and learning, as you work together. There is no limit to your potential as a people, if your evolution continues on a path that is truthful and beneficial to each other. One can do very improbable things in any time and place, at the expense of increasing the general surrounding disorder. Your post-Darwinian evolution is your

hope—but you must show your people that self-interest is other-interest. You must commit to a model of intelligence that is intrinsic, primordial, and pervasive, and stretch to it. You have that choice now, don't you?

S: Yes, as a whole people. But even if intelligence is universal, we need something to hold us together, to reach our potential. Our old religions did that in a way, holding up heaven and hell to individuals who needed moral direction—although I think the jury is out on whether they've done more harm or good. Anyway, how can we now make folks see individual benefit in social good?

    O: You have a better faith to offer, a rational faith, one that will last. It is a commitment, a trust, and an extension to truth and intelligent life, rather than to some miraculous or mystical revelation. Its hell is isolation. Its heaven is inclusion. You are given the capacity to love, to heal, to learn truth— and the freedom to choose them. That is all the sacred you can find—or will need. So, your people must re-think of themselves—not as isolated within their Flatland boundaries, but as participants in the larger reality of intelligent life. Isn't such a rational "faith" what you seek?

S: I guess that's the only place we could look. Anyway, it's the only source I would believe in. But there are still dying individuals and self-destructive societies. What shall I say to them?

    O: Say that entropy is real but it's not the whole story. Say there is more to them, more to relation, than is first evident to evolved senses. Say they can find self-interest in reaching for their "more" in others, that no faithful expression of intelligence, however small and brief it may seem, is lost, but integrated in the whole reality of intelligent life.

S: And what does your name mean?

O: I was a Flatlander. He was a Flatlander; she was a Flatlander. Owuza Flatlander.

S: So, I wuza Flatlander, too—?

     O: You were.

S: Will you talk to my most skeptical colleagues?

     O: Yes, with great respect. And then you and I will talk with your students.

BOOK II

With the skeptical scientist, January 2.

The Laboratory of Skeptical Science. *The study of biological life accelerates on Earth. In their laboratories, scientists are nearing the hinge of evolution, where it will be in their hands.*

I would not impose on your children—or on mine—a story that requires a supernatural or mystical explanation. As scientists and engineers familiar with energy and information theory, we know the reality of evolution and entropy. The story of Owuza Flatlander grows out of a commitment to truth. Our hope—that readers find in it comfort and direction—is based on rational thinking as the only dependable and lasting way there.

Flatland is about a way to live—a way that is both rational and hopeful. The story is, of course, a metaphor for the human condition. (All our language and art is necessarily metaphor, even mathematics.) Owuza Flatlander evolves, as we do, just far enough to survive in Flatland. He and his friends are missing something—they don't look up; they can't see off the page. But they learn how to live, and that adds a dimension to their experience.

The way to live is in commitment to truth and to intelligent life, that greatest of improbabilities. That is a logical choice, because it stretches our experience—assuming our interest is in freedom, communication, learning and building, healing, love, and beauty, all of which are enhanced by truthful and beneficial collaboration, and consistent with entropy reduction. Those are the experiences, the perceptions, that the Flatlanders learned to share.

The things we've held sacred are real. We can understand them. We can find a rational basis for beauty, and love, and for rationality itself: that, is, a commitment to truth. And we can find and understand a defensible moral compass based on that commitment: the reduction of ignorance and disintegration. These are the evidences of intelligence that we would accept wherever we detect them in the universe: the freedom and capacity to reduce entropy locally.

All expressions of intelligence (of which there are likely many in our universe and others) are free to make that choice, thus integrating. We would recognize intelligence anywhere by its ability to detect and choose order, and to exchange information, to communicate. From these grow other means of reducing entropy locally, the means that give intelligent life meaning: the ability to learn and build, to heal each other, to trust and make commitments and identify with each other (which we call love), to see and make beauty (that which reduces both informational and thermodynamic entropy).

The nature of those we love is in these expressions of intelligence. It is really those expressions, not the touch or the sight or the sound (which change in all of us over time), that we ultimately miss. And that is the true nature of ourselves. So, the Flatlanders find themselves and their missing friend in those lasting expressions of intelligence that they shared—still present and accessible in the truth and other expressions of intelligence that surround them.

Rebecca has drawn the Flatlanders and the intelligent universe around them all made of the same stuff, yet expressing diversity and the commitment to beneficial connection—to coherent integration of diverse experiences. Like a phased array with a widened aperture, when we work together we can resolve more of truth. (Untrue and non-cooperative efforts do not add coherently.) This is more than metaphor; it's the way the world works.

Of course, we cannot say with absolute certainty that anything is a "fact," even the simplest equations, because we're inside the system that we're trying to prove. ("Truth is a stronger notion than proof," to borrow a paraphrase of Gödel's Incompleteness Theorems.) Like the Flatlanders, we are far from perceiving all of reality. But we have the means to extend our experience beyond the boundaries of evolved senses. We have direct access to truth, as Penrose says, and to each other. These means are through the very evidences of intelligence that we have discovered.

All of experience is perception (but not all of reality, which we continue to explore). That extension to others, which psychologists call "identifying," is a rational element of what we call love, the other elements being commitment and trust. In exercising these with respect to others and to truth, we grow ourselves—our experience and our effect.

It is this "self" in which we're interested—the integration of those beneficial extensions, the truth we see through what seems an isolated individual. Consider the Stephen Hawking example: a body and brain could hardly be more isolated, yet his presence reaches all of us. Like the Flatlanders, his life doesn't "stop at the edges."

So, is there more to intelligence in our universe and others than just the sum of evolved brains? When all brains are asleep or disintegrated, is intelligence gone? To me, the logical answer is that intelligence is an intrinsic reality that does not need us to exist. The universe (or multiverse) is a logical, orderly place. Heisenberg called this the Central Order. I'll borrow that term, because it appeals to me as indicative of a whole reality, of which we have yet only partial understanding.

There is something, rather than nothing. That something is orderly. The universe(s) is probabilistic, but not capricious; that is, although many events are unpredictable (by us), they are not without reason, not without basis in physical and mathematical laws that we can count on. The physical universe in which we live, and probably other universes, depend on these laws, that order. Their birth, their initial inflation, their evolution, their space-time fabrics, are logical, not mystical.

It is unlikely that we are close to understanding all these laws.

But our capacity to seek and gradually figure them out is a wonderful expression of their Central Order. The species and mechanisms that evolve or are fabricated to evidence that expression are inconstant to the task, subject to entropy and sometimes accelerating its disintegrations and ignorance. But the potential of integration and learning is there, and its exercise gives us meaning.

Working together truthfully and to mutual benefit is in our self-interest as perceived individuals as well as collectively. It is integrative. The direction of that integration, which we are free to choose, is toward a Whole of which we can be part. That is a logical choice of a rational reality. Those of us engaged in reducing ignorance and disintegration see no limit to our learning, communicating, finding and healing others. That reach stretches ourselves as well as the common enterprise. Its goal is the Central Order, which is also what makes it possible: every such reach "expresses" the intrinsic reality of intelligence, and each expression is evidence of that reality. Although this experiment does not "prove" the hypotheses, it does test them.

The Central Order makes all reality possible. I think of it as a "field" of intelligence, which we express, rather like the Higgs field that our mass expresses, or an electromagnetic field expressed in our signal receptions.

Of course, some believe that nothing is real, and some believe that only the things we can measure or test are real, and some that only some pre-revealed creed is real; but these seem unlikely.

So, love and beauty, and progressive civilizations, have a rational base. As ways to live, these extend ourselves beyond the boundaries of isolated brains. If truthful and beneficial, they add coherently. That is how the Flatlanders learn to grow, escaping from their two-dimensional page.

The direction of that extension, that integration, is toward truth and intelligence. Although our perception of it is very limited, information grows without limit. Although individual expressions of intelligence appear (to locally-evolved senses) to be discrete and isolated, intelligence itself is without boundaries. We may say that all its expressions—those that are truthful and mutually beneficial—are "entangled" in the sense of being definable only as a single system.

Such expressions are without boundaries. They are our means of reducing entropy locally, and of sharing our perceptions—that is, our experience. While we thus express intelligence, we integrate toward a "Whole Each Other"—the full expression of relation and order. We express intelligence better together—not necessarily in physical proximity, but in reaching to each other via those means.

This is necessarily an asymptotic becoming, but it is a rational and hopeful commitment, a stretching of our experience of reality that gives us meaning.

Other "life forms" are probably more advanced than we in this stretching, and others less advanced, as we can see in the great myriad of life on Earth. Many forms likely evolve capable of expressing intelligence, and many more can be fabricated by those forms as they advance. Fred Hoyle envisioned a cloud of energetic particles that had developed the capacity to organize, learn, and choose direction. And, of course, all—including ourselves—vary enormously in their expression along our own "Flatlands."

But our interest is not in the forms, nor in their behavior when not expressing intelligence. Our interest is in their integrative capacity as they seek truth and beneficial connection. In that process, we can say that they are entangled in the single system of intelligence. In that process is our experience of what we call healing, and beauty, and love. That is where we place our trust, since intelligence is without boundaries.

So, we shall not say to our children that these familiar forms around them can ignore entropy, that the realities along our space-time surface can be indefinitely sustained or resurrected. But we can tell them that the integrative intelligence that they can choose to express has no boundaries; and that those who do express it truthfully, whatever their varying and perishable form, wherever they evolve, are entangled in that single system. As Rebecca has drawn in *Flatland*, we are in that expression all made of the same stuff.

Although our evolution has developed senses sufficient only to survive in Flatland, we do have logical means of expressing the nature of intelligence itself. As our children first look around, they see discrete packages of life, boundaries. Their perception, and the packages themselves, vary, come and go, wax and wane along the space-time surface that was born with our universe. That variation, that work of entropy, that loss, is real. But it's not the whole story. There is more to themselves, more to each other, than is apparent on first inspection with evolved senses. They may look instead for the continuity of intelligence among us, transcending these boundaries. They may look for ways to share the lasting nature of intelligent life, to find more of truth in collaboration. They have the means to stretch.

Seeking that truth, and being faithful to it as best we can find it, is a way to live. We have the freedom to communicate truthfully, to learn and teach, to build and heal and seek beauty, to extend our selves to others in commitment and trust. In those expressions of intelligence, we may add to the Whole of which we are part. We can sustain the coherence of our "entanglement" with others, with intelligence. That is a way to live.

As scientists familiar with entropy, we recognize this prescription as consistent with the reduction of ignorance and disorder; the resolution of farther truths with the wider aperture of diversity; and the coherent integration of truthful observations above the noise of error.

A Skeptical Dialogue

Sure, these abstract concepts reduce entropy locally—freedom, relation, communication, and so forth—but I have no evidence that they exist independently of our exercising them. When the person who was "transparent" to them dies or changes, they die with him. OK, someone else picks them up, but that doesn't help me.

In fact, I can quote you: "Beauty exists in the perception of intelligent beings..." So, there's no beauty after all the artists die, no love after all the lovers die.

Indeed, I am not arguing for the existence of abstract, entropy-reducing realities independent of intelligence. Quite the contrary: our freedom and capacity to see and make beauty—or to love, or organize—are capacities, and evidences, of intelligence itself.

What I do argue is for an intrinsic, primordial, pervasive intelligence that does not need our exercise to exist. And then I argue that our exercise of these capacities is a function of that intelligence, could not evolve or be constructed without it, and therefore should not be considered isolated from it. This is not an argument for "intelligent design" of our species, whose evolution is evidence-based, but simply for a universal and unifying potential for intelligent life.

I see no evidence to support either of those arguments.

That there is a non-capricious reality, that there are dependable laws, that there is truth in the multiverse, constitute for me clear evidence of a pervasive intelligence. We did not put it there. It is timeless, not a result of our evolution or manufacture.

And that we have direct access to truth and to the freedom and capacity of reducing entropy locally, constitute for me clear evidence of our manifestation of that intelligence.

I cannot prove you wrong about those evidences. But they do not establish our connection to some kind of primordial intelligence. I can create intelligent life in a petri dish. I can build intelligence into robots. I can make it, I can manipulate it, and then I can eliminate it.

Then you will have mimicked evolution and entropy, just at a greater speed. That is not creating or destroying intelligence itself. You built a mechanism capable of evidencing or exercising some of the capacities made possible by Intelligence. Evolution does the same thing, without intent and a lot slower, but with the same result. To the extent your child or robot expresses these capacities truthfully and beneficially, s/he adds to the Whole via integration. To the extent s/he expresses "relation," s/he is entangled in that Whole. You cannot subtract those additions from the Whole by killing the mechanism.

Those expressions come and go, age and die, wax and wane. There's no lasting personality.

Yes, we observe great variations in the exercise of intelligence as we can perceive them, because our observations are limited to our evolved senses and brain. As in Gödel's Incompleteness Theorems, we're stuck inside our own system of logic. Truth exists, however, independent of our proofs and despite different folks having different opinions of it; and so does intelligence.

No doubt many other mechanisms of intelligence exist throughout the Multiverse; and no doubt we'll eventually build brains. But each of these is limited to its own construct, and will be perceived within that system.

So, what's your theory of intelligence, if not just the coming and going of life-forms?

A Whole Intelligence would include all its expressions, without regard for time or space or their local boundaries. Different forms of intelligent life likely observe different characteristics of the Multiverse, using different sensors. We observe variations with our Flatland senses—beings come in and out of our perceived world, change and deteriorate over time, are assembled or destroyed. All these are the work of entropy and isolation; they are real within our perceptual limits, but if we remove those boundaries, relation and truth remain evident.

What is "truth"? And what is "intelligence"?

Truth is conformity to reality; that which passes every test, which prevails in every event. It is the integration of all information, which grows without limit. Information is a function of intelligence. Intelligence is the freedom and capacity to discover truth, to choose and bring about order, to share information. Truth and intelligence establish the Central Order from which all laws proceed and all experience derives. Truth and intelligence are timeless realities, pervasive and co-existing. These realities we discover; we don't invent them.

Where is your evidence for these postulates?

The evidence for truth is always and everywhere present, in things discovered, not invented. It is not just the sum of facts. It does not need evolution to exist.

The evidence for intelligence is always and everywhere present, in the freedom to organize and communicate. It is not just our ability to think. It does not need the evolution or construction of brains to exist.

The evidence for a Central Order, that which makes possible and universal the laws of physics, is always and everywhere present, in the constant and non-capricious operation of those laws. It is not just the local reduction of entropy. It does not need our organization to exist.

I can explain everything you claim without resorting to an intrinsic intelligence. The universe occurred by a random quantum fluctuation, an inflation of a quantum field. Probably other universes are doing the same. All I need is the laws of physics. And as to "Why is there something, instead of nothing?" I simply respond "Why not?"

Those laws of physics are consistent with Intelligence. The mathematics underlying them, and the application of the laws themselves, require the processing of information, which is a function of intelligence. It is much more credible to accept an intrinsic and inclusive Intelligence than to claim it arises only in discrete and isolated packages through an accidental evolution from nothing.

BOOK III

With the students, January 3.

So, what is that "more" of ourselves that we don't see with Flatland senses?

Our connection to the Whole of intelligent life is the "more." Intelligence is the "real thing"—it has no boundaries, in time or space. Flatland is a place of isolation, of boundaries, of "edges and endings." Rebecca has drawn our characters and the universe around them, however, all made of the "same stuff." The Flatlanders learn to express that "field" of intelligence that pervades all of reality—imperfectly of course, some of the time, subject to delusion and bias and so forth, but always with access to each other and to truth. That frees them from their Flatland edges.

But Owuza dies in your story, doesn't he? And then our "access" ends...

When his friends look only for the isolated blob of color they called Owuza, it is gone; it disintegrated according to the Flatland laws of entropy. When they search for the "more" that was really important to them, however—for Owuza's expression of love and learning and so forth—they find it still among them. The Flatlanders find a way to *live*. It is by expressing those timeless capacities of intelligent life truthfully and in mutual benefit. In that way, they can choose a "resonance" with the field of intelligence. In that expression, they are faithful to each other—and we might say to the "Whole Each Other," to all of intelligence, which is timeless.

The folks I loved, and the important things about them, aren't there any more in my life.

Are you so sure? If you look again at the Flatlanders when Owuza was not "there," what did they miss? The touch of a hand? The sound of a voice? The sight of his face (which changed all the time, as do yours and mine)? Those things do die, but they're not what Owuza's friends loved. They were confused because it seemed that freedom and learning, beauty and healing and loving were not there anymore. The sharing of those realities is really what they miss. Owuza brought them to those things, but that blob of color (in our case, "dust") did not *contain* them.

So, where is he?

Owuza was always in others, in the Whole of intelligent life. The "delusion," in Einstein's words, was the isolation of a bounded individual. Einstein said, "A human being is part of the whole...He experiences himself, his thoughts and feelings as something separated from the rest—a kind of optical delusion of his consciousness. This delusion is a kind of prison for us..." Rebecca's universe shows the unbounded reality of intelligent life, in which Owuza always lived. Although Flatland-evolved senses and brains and bodies die, the "more" of Owuza and his friends always lives, in that relation which is without boundaries. So, they find each other, and themselves, in the expression that "doesn't stop at the edges." It is that Whole in which they live.

But even our brains die, our ability to relate and express intelligence...

Yes, our brains evolved along with our senses, to survive in our Flatlands. They not only die, they sleep and can be manipulated, and dream and lose consciousness all the time. But they're not the whole story. They are not what we look for when we seek intelligent life around the universe. It is characterized by capacities that actually reduce disintegration locally: the ability to choose order and communicate about it. And those capacities make possible the "more" that the Flatlanders discover: learning and healing, making commitments and trusting, seeing beauty. Those choices reduce entropy (disintegration and ignorance) also; they reach to each other and to truth.

I can see that those "capacities" remain in the universe when a loved one dies. But doesn't her "capacity" die along with her brain?

Only if you consider that brain to contain her whole being. The Flatlanders learned to let go of that image. Owuza is not just an isolated, decaying end-product of Flatland evolution.

Life and death are Flatland things, as time is. They attach to isolated individuals. Flatland is full of edges and endings, and evolved senses do not see beyond them. They include the boundaries of those we name and treat as distinct, separate persons. To Flatland eyes, that person seems to contain "life," to be a source of intelligence. The evolved brain is indeed a wonderful processor of information, which you are gradually learning to augment and manipulate. No doubt there are many such processors in an infinite multiverse, some more advanced and some less. But each is isolated, subject to the time and entropy of its universe. It is not the container of what you love. Those expressions are found in relation. True and beneficial expressions of the universal intelligence are always included in it, and are therefore timeless. It is those expressions that you love.

But we seem limited to our "Flatland." How do we get in touch with that "more" of ourselves and those we loved?

The reality is, we are part of the Whole of intelligent life, as we express those capacities truthfully and in benefit to each other. That "Whole," that sharing of intelligence, makes relation possible. It's not limited to humans on planet Earth. We all "express" intelligent life, the way genes are expressed in our bodies. In that expression, we share intelligence with others—we're free to communicate with each other, heal each other, love each other. Those are ways that intelligence is expressed, by you and me and by those we can't see in Flatland.

So, is that "Whole Each Other" like God?

That word is too loaded for us. It means too many different things to different people. It often implies some supernatural being who performs miracles when he decides to. The Whole Intelligence we are talking about is real and rational, its expressions are natural, its evidences are how we'd detect intelligent life anywhere. But it has no boundaries.

That still sounds like what folks call "God."

The central reality we're talking about has many names. We have called it Intelligence, because we had to start with logic, with physical reality as we understand it and can test it—so the story would be grounded in truth. Heisenberg called it the "Central Order." Schrodinger and Einstein thought of a singular, universal consciousness. It includes all intelligent life, all of truth. Some might call it "God." By any name, we can trust the Whole Each Other, the full Intelligence that includes all faithful expressions, knowing that our full being extends there.

But people want a God who loves them, who takes care of them and those they've lost.

Yes, of course we want a human-like God who takes care of good people in an earth-like heaven after they die. It's an extension of what we're used to, and so it's comforting. But reality is better than that. It's not really the sight or sound or touch that we miss—those are the poor means that Flatland evolves for sensing love. Better means are available, as our characters in *Flatland* found. And they are lasting.

Does that Whole Intelligence, like, love us?

Yes, in a more real sense than romantic or otherwise perishable love in Flatland. It is a perfect commitment, trust, and extension, unconditionally given. That's what love is. The Whole Intelligence is perfectly *committed* to all its expressions, through the unconditional gift of freedom and our capacity to organize, communicate, learn, heal, love, and see beauty. We are perfectly *trusted*, in what the religious call "grace"—no matter what our past, we are able to be faithful today, and the gift of truth and each other is always available. Finally, the perfect *extension* of the Whole to us is our inclusion therein. So, we have this perfect, unconditional, permanent love. And we can love in return—through our commitment and extension to truth and life, our trust in that reality. That is what we mean by "faith."

I can see how helping each other expresses intelligence. But I still don't see how we can be part of each other.

By being part of the Whole, which includes all of "each other." The delusion that Einstein was talking about is our acceptance of boundaries. They're not the whole story. By reaching out to each other we reach to the Whole, and to our own Whole beings. By "each other" we imply a reciprocal extension, each in others. In a truthful and beneficial relation with all others, we can experience more or ourselves, more of the intelligence that we share. We think of that as the "Whole Each Other," by which we mean the complete inter-relation of all intelligent life that is truthful and inter-beneficial. In that infinitely-inclusive Whole we can put our faith.

But I can poison such relationships. We see every day people making destructive and untrue relationships.

Yes, intelligence is expressed also in freedom. That "field" of intelligence is the *potential* of relation. We can make destructive choices as well as constructive ones—or none at all. We can choose isolation or connection. We can choose to harm or to help. To the degree that our relationships build trust and commitment, to the degree that they promote healing and freedom and learning what's true, they "integrate coherently," as engineers say, in the direction of truth and mutual benefit. That is, our beneficial and truthful extensions reinforce each other, add up in phase, whereas untruthful and destructive choices interfere with each other, do not correlate.

How can we add to truth, if it already contains all information?

Because information grows without limit. We are free to add to truth and to discover it. We are also free to ignore it or deny it. Truthful expressions add coherently.

In recent history, we've seen lies and destruction take over whole nations with the cooperation of their people.

Adherence to either an untrue creed or an exclusive intent can always be chosen. But we seek a joining that is both true and inclusive. We can focus on that which enhances the life-qualities for all. In that focus, that choice, we grow toward the "more" of our beings.

Does the evolution of intelligent beings always follow that path?

Evolution is now in our hands. Its quantitative progress and its qualitative direction on Earth will henceforth be determined by this people. Probably many civilizations in our universe and others have reached this point and then blown themselves up or poisoned their nest or lost a microbial race. Probably others have advanced far beyond us, in learning and building, healing and freeing themselves of their Flatland boundaries. That would require sustained commitment to truth and each other.

So, is there any hope for humans on planet Earth?

Working together truthfully and to mutual benefit is in our self-interest as perceived individuals as well as collectively. It is integrative. The direction of that integration, which we are free to choose, is toward a Whole of which we can be part. That is a logical choice of a rational reality. Those of us engaged in reducing ignorance and disintegration see no limit to our learning, communicating, finding, and healing others. That reach stretches ourselves as well as the common enterprise. Its goal is the Central Order, which is also what makes it possible: every such reach "expresses" the intrinsic reality of intelligence, and each expression is evidence of that reality.

But is that Central Order benign? I mean, is there any good built into the universe? There seems to be plenty of evil.

Logically, we consider "good" to be that which benefits others, where "benefit" means enhancing their life-qualities—freedom, learning, healing, and so forth. Intelligence and the central order of the universe are what makes those capacities possible. They do not command it or necessitate their exercise. The expression of "good" or "evil" in any world lies entirely in the behavior of its intelligent inhabitants. Good in the world is a choice of truth and mutual benefit. Evil is the opposite choice. We can't define good people or bad people; only their behaviors and commitments are good or not.

Well, if good and evil are just what we make of the freedom and capacities we're given, what advantage does good have?

A shared commitment to the benefit of each other and to truth has the advantage of diversity and inclusion in that shared goal. We make more progress working together, integrating greater numbers and varied perceptions truthfully, than we can by exclusion and ignorance. We cannot know all of truth or reach all of each other, but we can commit to seeking both. That is a logical choice whose coherence makes it efficient.

Are you saying that these religious ideals are real?

Love and beauty, and progressive civilizations, have a rational base. As ways to live, these extend ourselves beyond the boundaries of isolated brains. The direction of that extension, that integration, is toward truth and intelligence. Although our perception of it is very limited, information grows without limit. Although individual expressions of intelligence appear (to locally-evolved senses) to be discrete and isolated, intelligence itself is without boundaries. We may say that all its faithful expressions are "entangled" in the sense of being definable only as a single system.

What help does that give me?

There's more to you than you're considering. In expressing intelligence faithfully, you live in the Whole of life and can add to it, just as a civilization progresses, as evolution develops qualitatively. It is such expressions that make up the Whole, in which you live. This is analogous to what the religious call the "love of God"—you have that perfect commitment, trust, and extension of Intelligence to all its faithful expressions. (By "faithful," we mean in commitment to truth and to the benefit of others.)

Such expressions are without "edges and endings," as we said in *Flatland*. They are our means of reducing entropy locally, and of sharing our perceptions—that is, our experience. While we express intelligence truthfully and beneficially, we integrate toward a "Whole Each Other"—the full expression of relation and order. We express intelligence better together—not necessarily in physical proximity, but in reaching to each other via those means.

But you said we're not equipped to perceive more of reality. Then how do we find it?

We can't find *all* of reality, because we're inside the puzzle. But each of us does have what the mathematician Roger Penrose called "direct access" to that objective reality, to truth. We also have access to what Scholar-Rabbi Martin Buber called "relation," to each other. Our access to truth is better from diverse viewpoints. We can find more of ourselves through each other, as our characters in *Flatland* learn to do. Being part of the Whole is the "more" of us.

How does that "access" work?

Every time you reach out in kindness, every time you stop to see beauty or make it, every time you set someone free or teach or learn or build together, every healing, every true communication, is a way to reduce entropy locally, a way to express intelligence. Those are the ways we can reach through our Flatland boundaries and help others do so, and in that sense they are sacred.

That inter-relation sounds good, and I can believe there's more to reality than we can see, but I could die tomorrow. Would I still be part of some Whole Intelligence?

What dies tomorrow is an isolated product of evolution. The Whole Intelligence does not die. We can choose to be part of that Whole—to add to it, to share deeply in the expressions of intelligent life. That kind of sharing builds relation on a lasting ground, one that does not respect time or place. It's like the quantum phenomenon of "entanglement," in which particles are related in such a way that their states cannot be described individually, but only as a single system.

You're not saying that we can become "entangled" like a couple of photons or something, are you?

No, as macro aggregations, we can't count on that experience because of what physicists call "decoherence"—the states of our pieces are all jumbled up by surrounding noise. But there is another reality that does not respect time or place: intelligence itself. So, we can experience an "entangled" relation with each other, and with the Whole Each Other, that is lasting, as the full expression of intelligent life.

How can we be sure there's more to us and our relations?

There is something, instead of nothing. Why? There is objective reality, an ultimate order to things. There is truth: some things are true and some are not. The basic laws of physics prevail, whether a "big bang" starts our universe or not. Why? Werner Heisenberg, discoverer of the Uncertainty Principle, called this the Central Order.

Detecting such order and choosing it is what intelligence does. Without that Central Order, that universal intelligence, there would be nothing.

What is "order," and why is that so important?

Order is a measure of improbability. Orderly things are unlikely. Disorder and ignorance are likely. It's a lot harder to organize and understand things than to let them fall apart. The fact that there are dependable physical laws, that there is a non-capricious reality, that there is truth at all, gives evidence of a pervasive intelligence.

So, where do we fit in?

What is the most improbable thing of all? We are. Intelligent life. We have "direct access to truth," in Roger Penrose's words. We have the freedom and capacity to do extremely improbable things, like communicating and healing and making commitments and seeing beauty. Those are the things our *Flatland* characters learned to do, allowing them to look up from their page and find a wider reality.

You mean we're part of a Central Order?

We mean there is in the Multiverse a Central Order—an intrinsic, timeless, pervasive, primordial Intelligence—to which each faithful expression of intelligent life is *transparent.* We love and admire those whose "transparency" improves our view of that central reality, through their expression of its evidences, its capacities. The Central Order, the infinite Intelligence, is always "now," as Erwin Schrödinger expressed it—always inclusive and complete, even though the local (evolved) means of expressing it, of seeing it, vary and come and go in their own "Flatlands."

Then is "transparency" an essential part of our identity?

Yes. What we see through those we love—and what we miss really, when they're gone from sight—is the view of the Central Order we saw through them. That's really the experience of love. But in their expression of its nature—of Truth and Intelligence—they were (and are) *part* of the Whole Each Other, adding to it. That Whole is all of the true and mutually-beneficial expressions of Intelligence.

It seems to me that our view of that "more" is hard to hold onto.

Yes, the many expressions of intelligence must be highly variable themselves throughout the Multiverse. They would have highly variable views of reality, as we do. Some are doubtless far more advanced than humans in their freedom and communication and learning and so forth, and others far more primitive.

In fact, that's true of us—sometimes we're fully awake and creative, sometimes unconscious or impaired. None of us is blameless, none succeeds at every try. But the view of the Central Order would be a single Whole—not a collection of isolated, temporary creatures falling victim to entropy or malice, but one system sharing the "state" of intelligence.

I guess most folks could believe that. But will it make them feel better?

Yes, it can. The lesson of *Flatland* is that relation has other dimensions, not limited to the space and time that were born with our universe. The experience of love is a realization of that extension. It's real because relation is grounded in Intelligence—as Erwin Schrödinger, the great quantum physicist, said, "always Now." We should not think of ourselves and those we love as isolated beings that can be separated by the effect of entropy on the senses. Each faithful expression we make is part of the Whole by definition: all together <u>are</u> the Whole, always complete but always growing by new expressions.

But we run out of time in our lives…

Time is a Flatland thing; it was born with the "big bang" of our universe. In Flatland, memories and commitments seem limited to our evolved brains, which are subject to decay and confusion. But

Intelligence itself has no such boundaries. In it, we have no edges or endings. This choice of joining among diverse parts is the real thing, not limited to its shadow on Flatland.

What is "intelligence"?

The universe is a logical place. It's not capricious, not without order. There is in the universe a freedom to choose order, to bring it about. That's what intelligence does. It reduces what scientists call "entropy"—ignorance and decay—wherever it's expressed. There are laws of physics that you can count on. All these things are evidences of intelligence as an intrinsic reality of being. If you were a SETI scientist, that's what you'd look for—evidence of a choice of order, an ability to exchange information, to learn. That gives intelligent beings the capacity to organize and build, to heal each other, to make commitments, to find beauty.

You keep saying "expressed." What do you mean by we "express" intelligence?

The way we express ourselves to others, the way mathematical expressions set out relationships, the way genes are expressed in our cells—the potential, the reality, the underlying truth is always there, but it is not always manifested or revealed, its effects are not always demonstrated. When a gene or an equation or a person is expressed, we experience its potential. Truth and intelligence are present in every event, but not always discovered or realized or manifested or acted upon—not always expressed.

Before my parents died, they forgot who I was, and no longer loved me.

Sure, all brains are subject to confusion, manipulation, decay, and so forth. So, it's important to realize that the person you love is not defined by, not isolated in, that shrunken mechanism. What you love is the sharing you experienced of intelligent life. Your inclusion, and theirs, in that Whole are timeless.

This still feels kind of cold and intellectual. Can we have a personal relationship with this Whole Each Other?

Our relationship is perfectly personal. More than the parent-in-heaven tradition. Immediate. Fully inclusive. Based in reality. In a love experience, we feel "part of each other." In our relationship with the whole beneficial and truthful expression of intelligent life—the Whole Each Other—we really *are*.

But that Whole Intelligence lets all kinds of evil and unfair things happen. That doesn't seem right.

Fortunately, we find no evidence of either divine intervention or a divine plan for the world. We can't blame God for the evil we do or the events of chance. Indeed, we would not want events dictated by some supernatural power whose purposes we cannot understand. We are free to set our own course, to learn and heal and love. As a people, we have the ability to steer evolution in a beneficial direction for all. Of course, that freedom also allows us to remain ignorant, destructive, or cruel. And, of course, things decay in our world, despite our best efforts. Our appeal is not for miracles but for our growth, our "becoming," our extension to more of reality.

Shouldn't such a "Whole Each Other" give us some help?

Yes, and we do have the greatest gift, that of truth and each other. By that we mean although the universe is probabilistic and subject to entropy, it is not capricious. Truth prevails in every event. Together we can grow in both understanding and healing. Our boundaries seem opaque to our evolved senses, but we are given the means to reach beyond them, in discovery and beauty and love.

What about what folks call a "religious experience"? Is it real?

When we're startled by something beautiful, or by a discovery or a healing or freedom, or by love, that's a connection to the Whole Intelligence. Of course it's real. We feel loved in those moments, by a Whole that includes all. We're embraced by the Intelligence that pervades all reality.

We try to be rational in every event, but we can drive around a mountain curve and see the morning sun on a snow-covered peak, and tears may come. In such moments, we might think "thus God sings to us." Of course we don't believe in an anthropomorphic God sending music from a heaven, but we do know that beauty reduces both forms of entropy, and that the infinite intelligence gives us the capacity to understand it. That is to us a sacred thing.

Is there any room for prayer in this logical explanation?

We're not against prayer, just against magic. Being rational just means being faithful to truth. A rational prayer would express thanks for the capacities of life, remorse for being unfaithful, and commitment to life and truth. That is a trust, a commitment to the benefit of others, an extension of ourselves to the Whole.

That's OK when I'm feeling strong. But when I'm grieving or sick or in pain, the world closes down around me and I can't think about commitments or gratitude.

That's exactly the Flatland experience. Those boundaries are not the whole story. Prayer in such times is realizing that the faithful expressions we and those we love have made are always included in a Whole that has no such boundaries. Our brains evolved to survive on planet Earth, so we're limited in what we can discern of a Whole Intelligence. But some things that cross these boundaries are within our experience, even when our isolated, individual capacities are shrunken by entropy—pieces of reality reaching beyond immediate sensory inputs. Beauty, love, and access to truth remain available.

How can I "realize" those things when I'm hurting so much?

You can focus on that "field" of intelligence that fills all reality, you and those you love. It crosses all boundaries of time and space and entropy. You can think of yourself in that field, free of those limits, resonant with that reality. Prayer is not asking for miracles, it's trusting our extension, our inclusion. You can address the Whole of which you're always part, knowing that these shrinking boundaries are not the whole story. You can love that Whole, just as you continue to love others even when in pain. That is, you can commit to the Whole of truth and life, and trust in them; and you can extend yourself to them. That is faith, and love.

How can we love something that's infinite, so beyond our reach?

Intelligence is infinite but it's certainly not beyond reach, because we're part of it. The reality of our life and our love is that they are made possible and sustained by Intelligence itself. Our expression of that Intelligence includes the capacity to love—to commit and trust and extend ourselves to others. Those capacities are a free gift, and in that gift we can feel loved. That is a more dependable love than any that relies on Flatland mechanisms.

I can't see how it's possible to communicate with an infinite intelligence— or why it should matter.

It does matter. We are free to join, to love, or not. That free choice is our communication to the Whole. Our message is our commitment, trust, and extension to truth and life. We make that decision every day. We're free to express that faith, or to withhold it.

How can we say anything new to a Whole Intelligence that already contains all information?

We know that information grows without limit—it's not "conserved," as the physicists say. So, although the Whole of truth is complete, it can always add new information. Alfred North Whitehead, the great mathematician-philosopher, put it this way: "God's conceptual nature is unchanged, by reason of its final completeness. But his derivative nature is consequent upon the creative advance of the world." If you substitute "truth" or "Intelligence" for "God" in Whitehead's concept, it's a perfectly rational description of this possibility.

But I want to talk to the one I've lost, and hear her answer.

Those who seem lost to Flatland eyes—not their form, but the faithful expressions they made with us—are always part of the Whole. They share the "listening." We are part of the Whole also, so we can receive messages whenever we listen. Those messages are what the Flatlanders found as expressions of Owuza's presence among them— the beauty and discovery and love that they share.

Can we put this in words, when we pray?

Of course. Words, and art, and mathematics, are all analogies of reality. We need them to focus our concepts. Some folks say "Thou," for example, when extending themselves to the Whole, because they grew up with that form of language reserved for the sacred. So, you might address "Thou in whom we live" when you pray. It's OK to use poetic or religious language and symbols if they help us perceive our real inclusion. It's OK to speak in words to one we love when she's not visible to Flatland eyes. Then really listen: watch for her and yourself in others, in those capacities she expressed faithfully. Reach out to all those expressing truthful and beneficial life. It is that Whole in which she and you live.

Well, trust and commitment are sort of like faith. But how can we extend ourselves to something that we're already a part of?

By growing in our expression of it. We can extend ourselves to the Whole Intelligence in which we live by extending to truth and life. Seeking truth and being truthful in all activities and thought, trusting in truth wherever it leads, and committing to truth in behavior and communication, constitute faithfulness to truth— which is to love truth. Same with life: committing to the benefit and preservation and growth of intelligent life, trusting the expressions of life that are found or encountered, extending one's self to others and their benefit, reducing ignorance and decay together, constitute faithfulness to life—which is to love life. Our highest understanding of truth and life is the realization of that Whole Intelligence in which we live.

I can hear folks saying, "I try that, but I just keep remembering, she's gone, she's gone!" Can you give them a less intellectual thought, something to hold on to?

Things seem "gone" because our senses can't see past Flatland, and our brains experience time. The *whole* person, however, and her relations, live in Intelligence, which is timeless. The sum of faithful expressions that you shared really is "there," just as she always was. As we start to realize that, we can actually experience it. What I perceive, I experience, I "feel." And when that perception is true, it is lasting.

But how do we perceive things that are beyond our reach?

You have to stop thinking of intelligent life as remote and other-worldly, and realize it is the closest reality. Actually, you do a lot of "seeing" without your eyes—for example, when you think about others, or solve a problem in your "head," or make a commitment, or conceive of something beautiful. That's what Roger Penrose meant by each of us having "direct access to truth." We have the capacity to perceive more than the immediate impulses of our five senses. Our expression of intelligence is what makes that possible.

To help perceive that reality, we might visualize a "field" of that intelligence in which we live—like gravity or a magnetic field, all around us and through us all the time, respecting no boundaries. That is the reality in which we live. And so do those we love.

That seems logical, but is it real? I mean, it's so beyond our experience...?

Actually, this is not unusual in nature. Everything extends

beyond its apparent edges. The reality of all things is that they overlap: their "fields" merge, some more strongly than others. It's really that extension that you miss, that sharing, that being like one, not the touch or sound or sight, which changed all the time.

Well, I can believe that of micro particles, but people...?

If you accept the reality of the Whole, shared Intelligence, you can see that your own full self extends without limit. All of our faithful expressions are part of that Whole life that we share, that sustains us. Of course, you can't see this with Flatland eyes. But you can find manifestations of this Intelligence in its evidences around you. Your Flatland capacities are always limited and variable—you don't see your freedom, you don't always discover, or heal. But you can seek your wider self, your full extent, always in others and in truth.

Can I really go there, and find someone I've lost?

You're already there. You don't have to die or be transformed or go to heaven. You don't have to discover something new, just see what you've been missing. We're talking about a timeless reality that includes all. So, think of her as next to you in your full being, sharing life, free of Flatland edges. Your memories of truthful and life-expressing events like love remain real, because they are timeless. Enter that space in thought. Remember those expressions of her nature that you really want to recover—those that expressed her true being and relation to you. Those that were true and loving live on in that "space" you share, in others' true expressions, in the Intelligence that has no boundaries.

That all seems logical, but how can we feel the presence of those we've lost, when we can't see or touch them?

How did you feel her presence when she was in another room?

Well, I knew she was there, I knew I could go see her, I knew she would come back.

Exactly. It's that "knowing" that intelligence does. It dissolves the boundaries that isolate us.

But it doesn't help to make believe, or just imagine...

The thing to believe in is reality. We can trust in what's real, we don't have to make it up. The *Flatland* lesson is that reality is not limited to what we can see and touch.

OK, I know there's a reality out there that I don't see or touch. I can believe she's part of it. But it doesn't help if I can't go there. I know you said we're already "there" and together and so forth, but I don't have any evidence of that. The only evidence I have is of our bodies disintegrating.

And that evidence is accurate—our bodies do disintegrate, all things in our universe experience entropy. But that's not the *whole* reality. We have all around us the evidences of intelligent life reducing entropy, wherever that life is expressed. Those are the evidences our characters in *Flatland* found.

What about a dying baby, who never had time to choose relation or learn truth?

Our perception of that child's expression of intelligence is first via perishable means. We see little arms reaching out to be held, feel them around our neck, hear the gurgling of a contented baby, see the smile aimed at our face, breathe in the infant smells. Those senses are subject to entropy. So, as we sadly know, are the little arms, the smile, the sounds, the touch. Tennyson says that for us all: "Oh for the touch of a vanished hand, and the sound of a voice that is still."

Our senses and brains are bound by space and time, and so are the physical evidences that they read. Those things deteriorate, can be confused and manipulated and attacked and lost. But they are not the whole story.

The child gave us intense evidence of a reality that does not need our senses or our brains to exist. Those evidences were her expression of communication, trust, and extension: manifestations of intelligence. Our response, beyond our own touching and hearing and seeing, was in commitment, extension, healing, recognition of beauty, and communication, freely given. Those—not the smells and sounds and touches—are the evidences of a timeless reality, the capacities we all derive from the Whole Intelligence.

But those evidences are hard to see in grief.

Our recognition of another's being seems at first to be limited to what we can discern via five senses evolved on planet Earth. But we are given a capacity and a freedom to recognize reality by better means. We have the Penrose "direct access to truth," and the Buber "relation." These support a reach through Flatland boundaries. They support a perception, and thus an experience, of intelligence itself. What we perceive by those better means is a glimpse of the Whole, the Intelligence in which we live, and the relation it sustains. In a true perception, we recognize a true expression of that reality.

It does not matter how brief our recognition or how flawed our discernment. The Whole is undiminished by our loss of sight.

Is that like a "soul," that continues when the body dies?

As we said about heaven, reality is better than that. We're not looking for an isolated spirit somehow freed of physical laws, taking flight from the body when it dies. Our condition is now and real. Rebecca has drawn all the Flatlanders—all expressions of life—made of the "same stuff," always included in the Whole Intelligence. To locally-evolved senses, birth and death seem like beginnings and endings of life, temporary and constrained. But Intelligence itself is timeless and boundless.

So, if we're all part of the same stuff, is that like a big soup that dissolves us all?

No, one of the evidences of intelligent life is its ability to choose, its freedom. There are probably an infinite number of what we call "individuals" in our universe and others—wildly miscellaneous, to borrow a Jane Langton phrase—each of whom can choose to join or to remain isolated. And we know that our grasp of reality, of truth, increases with diversity, with many different points of view. The nature of truth is infinite diversity, not homogeneity. In joining we shed our Flatland boundaries but retain our freedom. We're not hoping for a continuity of the isolated individual, nor for absorption into a homogeneous will, but for a wider identity that more fully expresses the life-capacities, including freedom.

What if I build a robot, and it learns to think and communicate and build things? Will it be made of the same stuff?

We're already getting close to making human-like robots, and building self-learning programs into them. But this is not really different from deciding to conceive a child—organic creatures seem more "natural," but we can make them in a lab and we can take them apart, just like a robot. Someday soon we'll be able to engineer chromosomal changes, replace cells, and extend the function of our bodies and even our brains indefinitely. There's nothing sacred about DNA evolution. The important point is that the primordial Intelligence pervades all of reality, and is surely expressed in infinitely different forms in our universe and others, however imperfectly and sporadically in each case. The robot and the baby and other life forms give evidence of this pervasive reality, but they don't create some kind of new "soul."

So, I'm no better than a robot?

"Better" does not attach to individuals, but to behavior and commitments. While we're awake and well, we can choose greater or lesser expression of the life-evidences made possible by the "field" of intelligence in which we live. To varying degrees, we can love and see beauty, we can heal and learn and communicate; those capacities

vary, but we can always be faithful to them. That choice is our identity, how we are truly known and remembered. None of us "is" good,
but all who can make that choice can express good.

*Well, if I can't see all of me with my Flatland senses, is the rest of me
making the same choices?*

That question is still seeking an isolated, individual "soul." It is
in relation with the Whole Each Other that we live. That supreme
relation, to truth and intelligent life, is always intact. We perceive
it in varying degrees. We can improve our perception through
faithfulness.

So, my natural being is shared with others?

Our natural condition is inclusion in the Whole; expression
of the pervasive "field" of Intelligence; entanglement with others
who express intelligence beneficially and truthfully. Surely there is
an infinite variety of intelligent beings in our universe and others,
some of which we would not recognize as "organic," some we'd call
"mechanical," some perhaps like the cloud of energetic particles
imagined by Fred Hoyle is his story of *The Black Cloud.* Some being
born or assembled, some disintegrating. The expression of intelligence is a matter of degree; it's not absolute or continuous in any
isolated form. But it includes an element of choice, to be truthful and
beneficial or to be isolated.

*You mean there's no difference between living things and non-living
things?*

We mean the difference is indistinct and variable regarding how
we express intelligence. We see that in our own experience. Think of
the evidences of intelligent life and to what extent they are expressed.
Can you say exactly which creature is free to choose order, to communicate, to make commitments—or when? Is the "brain-dead" person
alive? The 4-cell or 8-cell fetus? A bacterium? Intelligence is seen not

only in many forms but in many degrees. Remember we're talking about *intelligent* life, not some fuzzy organic definition. There's no absolute on or off in our expression of intelligence. Only in isolation are there edges and endings. In making up part of the Whole Intelligence, we live in relation.

Can I believe that our familiar bodies and brains are somehow unnatural?

Our unnatural condition is isolation within Flatland boundaries. Our perception of what's "natural" can be lost, rather like the way the entanglement of elemental particles is lost in what physicists call "decoherence," when surrounding conditions change their state in unpredictable ways. We experience boundaries through the limited senses evolved to survive in our world. But we can invest in "becoming" as a whole people.

I can see that, but I do experience what you call "boundaries," all the time.

Yes, we do. All of experience is perception—but not all of reality. Through our perception of each full expression, we can experience the Whole. It does not matter how that expression was constructed, organically or mechanically or genetically modified or by some process we don't yet understand. Our own true and beneficial expression, and that of those we love, and of all others, make up the Whole Each Other. We—the infinite "we" of all truthful, mutually-beneficial, intelligent life—<u>are</u> the Whole.

But the way I perceive that is often weak, and can die.

Exactly so. The mechanics by which we perceive ourselves and others are subject to entropy, to malice, to chance. Our brains deteriorate and die, along with our bodies and senses. But they are not all of who we are. The isolated creature dies, whether evolved or constructed, whether on our planet or others. Our inclusion in the Whole of life does not. What we've called "relation" rests in our mutual inclusion. It is not dependent on such mechanics.

How can that help me if I'm losing the power to think? When I'm grieving or sick or in pain, I can't go through an intellectual exercise to feel in touch with truth. I just feel despair.

That's why we wrote *Flatland*. We all need a way to "internalize" a sense of freedom from our Flatland boundaries. We have to "feel" our unbreakable entanglement with those we love. We have to really *perceive* our inclusion in the primordial Intelligence. Rebecca's images can evoke those perceptions. We as adults can look at them and sense a kind of embrace—just as the child can sense reality before s/he can put it into words. When the world seems to close down around us, we need a real sense of connection, of inclusion. *Flatland* is a story of entanglement that doesn't break, of no boundaries, of an infinite ground of our being. And those things are real.

Where can I see those three realities in her drawings?

Look how she has drawn our characters and the universe around them all made of the "same stuff," entangled forever. Look at the characters escaping from their Flatland boundaries. Look at the infinite sky, the Whole Each Other that the Flatlanders finally see when they "look up."

So how do we "look up" in our grief?

Think of that life-giving reality as all around you and through you and through those you love, without edges or endings. We tried to convey that idea with Rebecca's sky, the way it includes all life. A field of gravity is like that, or a magnetic field, a TV broadcast or cell-phone signal, or the Higgs field that gives all things mass.

Well, I suppose it's impossible to capture all that in words, and so you combined them with art.

Yes, once we're sure of the "rigor" of our reasoning, we can make use of other routes to truth. Poetry and art, for example, help us see connections. They evoke perceptions from our deepest experiences, some out of reach of declarative prose. Metaphor and allegory help us look at a question from different angles; like diversity, they collect different viewpoints. Sometimes religious expressions can challenge us to look above our basic incentives. Beauty and discovery in all their forms can startle us into a sense of that embrace—we've all had that "religious experience." We can go there, even in our darkest despair.

But all of these have to be faithful to truth, wherever it leads. They have to point to what is real, not just to what we wish. Each of us has, as Roger Penrose says, "direct access to truth," to Platonic reality. Penrose uses the community of mathematicians as his example, and "pure mathematical fact" as the fragment of truth they share, but the principle is broader. When beauty or metaphor or religion helps our perception of the real, they can be faithful to truth, just as mathematics can.

All language—even the language of mathematics—is analogy. Since we're inside the puzzle, we can't see all of truth objectively, but must do our best to reason outward faithfully. Analogy and metaphor are necessary means.

Can you give us another good analogy or metaphor to hang onto?

Think of a field that stretches out forever, with a high fence cutting across it. You can't see over the fence, but that does not limit the field. The field is a metaphor of infinite intelligence. All intelligent life is rooted in that infinite field. Those roots entangle through love, and that entanglement is not bounded by the fences. Various expressions of life have their own "fences"—their own time and spatial limits—which are real but not the whole story. Ideas and beauty soar over the fences. Although the picture is metaphor, those concepts are real.

Now I have to ask if that experience would be real, or just a psychological trick?

All of experience is perception, but not all of reality. By that we mean that each of us perceives only that brought in to his or her brain by his five senses, and by what she can figure out or remember or discover. That is the entire experience of the isolated individual. It is clearly not all that's real. The illusion, as Einstein says, is the isolation. So, each of us can begin to experience the wider reality by perceiving our inclusion the Whole. How? By doing what the Flatlanders did: set someone free, communicate, learn and teach, heal others, love, see and make beauty. Those are the evidences of intelligent life, the life-expressions that reduce ignorance and decay. We find our wider selves and our "lost" relations in those expressions, through others. The things we hold sacred are real, after all.

But maybe "intelligence" is just in brains or computers or other forms…?

The expressions of intelligent life flicker in and out, in all of us, mixed with illusions and sleeping and illness and all manner of delusions. But it's not hard to recognize. Its evidences are those capacities of seeing order and choosing it, communicating, learning, healing, and so forth. We can stretch ourselves by exercising those capacities. And we can recognize them in others, thus reaching toward the Whole of which we're part. The illusion is the isolated identity, confined to the individual Flatland mechanisms. But we can look for life in others' expressions, and share it. Every encounter is an opportunity to perceive that reality in others, and to offer it to them, and thus to find the more of ourselves together. And to see the missing one.

But it's kind of hard to see these good things in a lot of people out there, unattractive, selfish, hurting, ignorant…isn't it?

Every bad behavior is mirrored in our own. None of us is admirable all the time. But all of us can express those lasting capacities some of the time. That's what we're looking for—just occasional evidences of the faithful parts of the missing person's nature. They can be glimpsed in the unlikeliest of places. The laws of physics let us pump good winter heat out of a cold lake, or a tiny signal out of a noisy broadcast. When we accept the bad in each of us, we can understand the good in all of us.

I read there's a pre-wired part of our brains that evolved to survive the realization that humans have, that we'll all die, and that's why we believe in God, in transcendent things…?

Of course—we're pre-wired for all kinds of instinctual behaviors and ways of thinking, based on our Flatland survival pressures. Not only that, our brains can be manipulated by electrical or chemical means, or by suggestion, and we can even some day make a brain. Sure, we want to be the center of the universe, we want the world to be flat…But we can reason logically and choose rationally. Wanting something to be true doesn't make it false. Wanting something to be false doesn't make it true. Our predispositions do not explain our findings. The genius of intelligence is its capacity to reason, to test. Truth is that which prevails in every event, which passes every test, which contains all information. We can't do all that, but our expression of Intelligence gives us the ability to reason beyond instinct. And, of course, we see farther and more clearly together, using our diverse viewpoints.

So, is this a new religion?

No, the Flatland metaphor is consistent with any true religion we know. We have tried to be respectful of all religious beliefs. The story deals with seeking truth every way we can, and how little of the Whole Reality can be seen with Flatland eyes.

You've quoted a lot of scientists but no prophets. Don't you come down on the side of science, instead of religion?

Our conviction is that both science and religion are seeking truth. There should be no conflict. There are of course scientists who don't believe there is any objective reality beyond what they can measure, and there are religious folk who believe that all that needs knowing is already revealed; we don't agree with either, but we are not trying to prove them wrong.

But if you're trying to agree with all religions and all sciences, what can you say?

We're not attempting either a scientific breakthrough or a religious revelation. We're simply offering a way to look at some self-evident realities and how they're connected. In that expanded view and relation, we hope our readers may find comfort and meaning and direction.

*The following is a personal essay
from the author.*

Diversity and Life

My grandchildren stand at the hinge of human evolution, now for the first time in the hands of their generation. Samantha and Abigail are young scientists squaring their technical discoveries with their experience of love and beauty. They study the farthest things in our universe, and the nearest—stars racing fastest away, and our own brains. I trust Samantha and Abigail, and commend to them this search.

They are not alone. Their missions are the same, just at different scales—to strengthen the exchange of information, in commitment to truth and to mutual benefit.

So . . . there is a fuzzy nebula at the farthest edge of the observable universe. Perhaps it is an undocumented galaxy. To see it more clearly, Samantha solicits observations from many different, widely separated telescopes of varying design, some on Earth and some in space. These observers see different views of the nebula, at different times and wavelengths, from different angles. Samantha wants to combine their observations, to see more than she could see alone.

Now the observers have a choice. Each could decide he is seeing a unique object, and give it a name, and argue with the other observers about whose description is right, or who gets credit for the discovery. Or he could cooperate with the other observers and with Samantha, to combine their insights into a greater perception, and share it.

Samantha suggests that each observer's self-interest can best be found in their collective interest. Diversity is their strength, because it opens a wider view, contributes a greater range of perceptions.

But Samantha must combine these perceptions "coherently"—in phase, capturing the maximum available information in their sum. Her processor is designed to do this integration constructively, so long as the inputs she receives from her collaborators are truthful to the best of their ability and contributed for mutual benefit. (Inputs can combine destructively also, if they are not brought in phase, or are untruthful, or used more for selfish than mutual benefit.)

Abigail's experience is similar, in her study of the brain. Countless neurons must "fire" coherently in the same way, contributing together to constructive action. They can also combine destructively if out of phase or impeded or randomly triggered. Abigail understands the wonderful complexity of this process, and its potential for creativity, or destruction. She also understands that it does not contain the whole person, who must be seen in relation to others.

Nor do these observations contain the whole truth itself. The galaxy exists without our observing it. Intelligence exists without our evolving brains to express it. Truth exists of itself. But all observers have access to the same central reality. They can choose truth or conjecture. They can choose integration or isolation. They have this great privilege of choice.

There is an intrinsic reality to these faithful expressions of intelligence, not limited to, not contained in, the mechanics of their expression. It is not the telescopes that integrate, nor their operators, nor the neurons in our brains, but the truthful information they share, driven by the intent of mutual benefit.

So these young scientists understand coherent integration across the whole scale of human experience. They can add to the common understanding, to the whole experience of observers everywhere.

And that is our goal—to integrate this people coherently, that is, truthfully and to mutual benefit—to heal their separation, so they need not suffer isolation, but experience the sharing of intelligent life. That is the means of connection to Heisenberg's "Central Order."

Samantha is a writer, and recalls Shakespeare's view of this Order:

> *. . . Look how the floor of heaven*
> *Is thick inlaid with patines of bright gold:*
> *There's not the smallest orb which thou behold'st*
> *But in his motion like an angel sings . . .*

Shakespeare had in mind the "music of the spheres" that ancient astronomers saw in the apparent motion of heavenly objects. We understand that motion now, in ways that could not be grasped in Shakespeare's time. But it is no less beautiful. The universal order that makes it possible sustains us all.

My young scientists are devoted not just to truth, but to each other—to the widest "each other" they can find. Their study of the farthest things and the nearest things is strengthening that devotion. They can find in their work the integration of diversity and the discovery of order—a rational faith in the highest expression of intelligent life.

This is my bequest to them.

Something There Is

There is something, rather than nothing. That something is orderly; it is not capricious, not magic, not mystical. There are universal laws one can count on. Heisenberg called this reality the "Central Order."

The achievement of order requires intelligence, the processing of information, the reduction of entropy. That reality exists of itself. It has no boundaries in time or space, as the basic laws do not, as the Central Order does not.

Those universal laws, that central order, would support many kinds of intelligence. It is highly unlikely that ours is the only one. It could be that we are surrounded by many expressions of intelligence that are simply not detectable by the senses evolved to survive on

Planet Earth, or by the extensions we have contrived, using the media familiar to us. We are surely surrounded by Heisenberg's "Central Order" and the potential of intelligence.

How would such an intelligence be expressed around the multiverse, through all of reality? Not the forms or the mechanisms, but the capacities themselves that intelligence makes possible—what capabilities can be exercised because of that pervasive reality? What would constitute evidence of intelligence, anywhere in an infinite multiverse? What should the SETI scientist be listening for?

Senses and brains that evolved through natural selection to survive in any particular world would not be equipped to perceive all of reality. But because of pervasive intelligence, any evolutionary process might develop partial capacities to reduce disintegration and ignorance—to reduce entropy—locally. Evidences of such capacities would include the ability to recognize order, the freedom to choose it, and the capacity to communicate about it.

From these basic capacities, other capabilities of reducing entropy would likely evolve: learning, organizing, healing. And then the ability and proclivity to make commitments and trust and extend to each other, essential to survival. (We call that special ability "love.")

In each isolated world, those developing such capabilities or fabricating them would consider their achievement unique to their species, and their view of reality complete. We call each such world, with its self-admiration, a "Flatland."

Nevertheless, if such expressions, however evolved and exercised, choose to be truthful and mutually beneficial, they could add "coherently," integrating in the whole reality of intelligent life. Expressing in part the whole field of intelligence, we might expect them to be "entangled" in the quantum-mechanical sense; that is, they could be described only as a single system, not in isolation. Evolved senses do not observe this entanglement, because of the decoherence introduced by their limitations. The whole system of intelligence, however, would be without boundaries, and its integrative expressions components of a boundless reality.

So although each universe is subject to the laws of entropy, the local capacity to reduce entropy (by this choice of integration) is sustained by the Central Order—which is of course timeless, because "time" is a phenomenon of each universe.

It is that reality of integrating expressions in which we are interested. In each universe, with widely variant evolved means of observing, their full reality would be observed incompletely. They may be given names; Shelley, for example.

Such names may be attached to forms and means that are limited in their own space-time frameworks. But our interest is not in the forms or the means, but in the actual expressions of the coherent, integrating reality.

Consider the electron, for example. It is completely defined by its spin, charge, and mass. It *is* those expressions of fundamental fields in nature. They're not qualities of something else: there is nothing else to the electron.[1] Now consider what you love in another. It is her truthful and beneficial expressions of the field of intelligence. Those are not qualities contained in a three-dimensional glob. Like the electron, they *are* the reality that you love.

So we can say this to our children, and to those bereaved by any loss: When you look for one you love—or for your self—look for those expressions in others, in the "whole each other," as far as you can see it. They do not always express them, nor do you. It is the truthful and mutually-beneficial expressions, however brief and limited their exercise in Flatland, that integrate in the Whole and become one.

Life and death are Flatland things, because time is. Your choice is between integration and isolation.

Thus you, and those you love, have your being together in that whole reality. None of us expresses that all the time, in every event.

---

1. Sayre, *Something There Is*, Peter E. Randall Publisher, 2014; pp. 217–218.

It is not the Flatland "you" that so lives. Nor is it an untruthful, non-beneficial behavior and experience.

The exercise of truthful and beneficial relation is what you really love in others. It is that exercise that integrates coherently in the Whole, that adds to it. The relation among such expressions is a sharing of those unbounded realities. This integration is the choice of extension to others.

That Whole therefore includes us—the "us" that expresses healing and love and learning faithfully, the "us" that integrates, adding to a complete reality that has no boundaries.

So are we loved by such a whole reality? If we see "love" as commitment, trust, and extension to each other, that Whole Each Other loves us perfectly. Our evolved perceptions equate the expression of love with an entity and an experience isolated in time and space. But no individual is the source of love. It is derivative of a whole reality. These expressions are not entire of themselves, are not complete. They come and go in our experience, but the Whole that they evidence remains. In that Whole we live, and are loved.

About the Author

David Sayre is the author of the essay sequence *Something There Is: Seeking a Rational Faith for Our Children* (Peter E. Randall, Publisher, 2014); the mystery novel *The Great Improbability* (Randall, 2010); and the children's picture book *Flatland*, illustrated by Rebecca Emberley and recommended by Parents' Choice (Two Little Birds Books, 2014).

Sayre is an engineer, entrepreneur, and writer, the father of five children and five businesses. His inventions and companies have changed the way we use energy and communication—those two sciences that contend with "entropy," which measures both disintegration and ignorance. Sayre's nonprofit organizations have also contended with entropy, in the human struggles with criminal justice, job creation, and mental health.

Finding intelligent life beyond our Flatland is all about reducing entropy. In his autobiographical *Something There Is*, Sayre described the progress made in that search by our greatest scientists. But suddenly evolution itself is in our hands. Sayre's *Dialogues* confront that existential challenge.

Visit davidsayre.com to learn more about the author and his work.